Reduce Your

Engineering Drawing Errors

REDUCE YOUR ENGINEERING DRAWING ERRORS

Preventing the Most Common Mistakes

Ronald E. Hanifan

Reduce Your Engineering Drawing Errors: Preventing the Most Common Mistakes
Copyright © Momentum Press©, LLC, 2010

First published in 2010 by
Momentum Press©, LLC
222 East 46th Street, New York, N.Y. 10017
www.momentumpress.net

ISBN-13: 978-1-60650-210-5   (softcover)
ISBN-10: 1-60650-210-7   (softcover)

ISBN-13: 978-1-60650-211-2   (e-book)
ISBN-10: 1-60650-211-5   (e-book)

DOI forthcoming

Cover Design by Jonathan Pennell
Interior Design by J. K. Eckert & Co., Inc.

First Edition January 2010

10 9 8 7 6 5 4 3 2 1

Printed in Taiwan

CONTENTS

PREFACE

In this book, I will discuss only the most common errors that appear on engineering drawings and the basic usage and understanding of the most frequently used drawings.

All drawings will contain errors, but if you can eliminate many of those errors before the engineering design checker or your supervisor reviews your drawing, it will go through much easier. Your reputation is at stake! Your supervisor and the engineering design checker will see everyone's work and know their errors. They know your weak areas and who produces good work and who doesn't. It is helpful to know what they look for—or should be looking for.

Herein we discuss the most common errors on drawings and the most frequently used drawing types. The data contained on these drawings, when not done properly, is subject to interpretation instead of universally published rules. Geometric dimensioning and tolerancing has become very complex, but it need not be. The best policy is to keep it simple but stringent enough to build good parts. As the drawing becomes more complex, the probability of misinterpretation increases. Also, it can increase the cost of the product and inspection tooling. Engineering is a field where rules are structured and uniform to allow for global repeatability in interpretation. These rules are the foundation of our profession and should never be violated, regardless how minor. The impact created from these violations may range from minor to major, immediate or future. Procurement, manufacture, inspection, stocking, and assembly and sparing of parts or equipment may be impaired. There normally is no real reason to deviate from established and approved rules.

1

FIRST STEP IS DONE, DESIGN IS COMPLETE

You have labored many hours developing your design, and now you are finally done. Whichever computer aided design (CAD) system you used, it has been a proven to be a tremendous asset, and you are very proud of your accomplishment. However, now it is time to prepare drawings so that parts can be fabricated, inspected, subcontracted, assembled, tested, or procured. The initial design is a very important part of your effort, but now it is time for the real world and the documentation of your product. This is the most important step in your design. This is what everyone will see, and it is what people will use to evaluate you and your accomplishments. The drawing is the one item that is visible to everyone. The primary focus is to document your design on drawings with such accuracy that no bad parts will be made and to ensure that there is no ambiguous information. The reality is that every drawing will contain errors; some will be serious and some minor. Even seemingly minor mistakes can cost millions of dollars. Regardless of the type of error, someone will be impacted.

From your drawing, procurement will order materials/parts or will subcontract parts to be made, production will set in motion the assembly steps, parts will be made, inspection plans will be developed, and drawings will be sent out to prospective manufacturers. All drawings will contain mistakes, but with careful attention, the severity of these mistakes can be minimized. Everyone will scrutinize your drawing, and your reputation, your department, and the company are at risk.

What you designed, as shown in your CAD system, is designed at nominal (average), and everything fits. It is perfect! But in reality, when parts are fabricated, they will contain many errors because of tool wear, operator errors, or misinterpretations. The drawing that you have prepared defines how much error is acceptable.

The data on your tube normally goes nowhere; it is the paper (drawing) that will now receive everyone's attention. Manufacturing, procurement, and inspection will be the main users of the data that you have developed. Drawings will be sent out for quotes to fabricate or buy, and parts will be inspected, manufactured, and assembled from the information contained on these drawings. There are instances in which a model is sent to subcontractors in lieu of a drawing; however, this very seldom works.

The information on the drawing is to be presented in such a manner that there can be no misinterpretation of the intent of the end item. All features and information require complete inspection criteria (acceptance/rejection). This sounds easy; however, it isn't. Each and every part of your design has to be documented in some manner. It may consist of new drawings or usage of existing drawings, commercial specifications, vendor parts, and so on. Materials and finishes need to be specified, configuration completely defined, and any other necessary characteristics such as installation of inserts, burr removal, flash specified, parting lines, surface texture, and so forth all need to be specified.

Requirements vary, depending on the type of item being depicted. It is very important to keep in mind the intended uses of the drawing. It may have to meet requirements imposed by a contract or your own company's standards. The drawing also becomes a legal and binding contract between your company and other users. Primarily, your drawing will be used by a manufacturer, but it is not a manufacturing drawing. It is an end-item drawing, defining only end-item requirements and not the methods of accomplishing the action. Hence, only a hole size is given, and not the method of manufacture, unless the actual method of manufacture is critical. Remember—a method is very difficult to inspect, and the method, when specified, does become an item requiring inspection.

You have dimensioned your parts completely, and, if done properly, inspection will have no problem understanding your intent. In all probability, you used geometric dimensioning and tolerancing (GD&T) to define your product. GD&T is a very complex endeavor, and you must understand exactly what you have placed on your drawing. Understanding of your tolerance zone, the shape and extent of it, and how the size of it may increase/decrease is of the utmost importance. The more exotic your GD&T application is, the greater the difficulty in the interpretation. If you don't understand completely the tolerance zone, then you can't expect others to understand. If the application is not contained in ANSI/ASME Y14.5, then you cannot use it. ANSI/ASME Y14.5 *contains the only approved rules,* and any other documents contain only someone else's interpretations. The best rule of thumb is to keep it simple but stringent enough to build good parts. Remember, the tolerances increase the difficulty of fabrication and inspection, which impacts the cost.

It is very important to pick the right type of drawing for the intended function. There are detail (part) drawings, assembly drawings, testing specifications, many different types of procurement drawings, drawings for altering existing parts, reference drawings (such as piping documents), schematics, interconnect diagrams, installation drawings, and others. Assembly drawings will contain a Parts List or Bill of Material (separate or integral). The most common of these drawings, and their requirements, will be discussed in Chapter 4. These drawings and their requirements are controlled by ASME Y14.24 and ASME Y14.100. ASME Y14.24 specifies the types of drawings allowed and their requirements, whereas ASME Y.14.100 specifies approved engineering drawing practices.

2

UNDERSTAND THE DATA THAT YOU HAVE PLACED ON THE DRAWING

Drawings and the errors that appear on drawings haven't really changed in over 50 years. The only thing that has changed is the manner of drawing preparation and the dimensioning practices. *Dimensioning and tolerancing* has been expanded by use of a more advanced *geometric dimensioning and tolerancing* to clarify the intent. Geometric dimensioning and tolerancing has been around and widely used since 1960, but the applications were much simpler and easier to understand back then. It is imperative that the designer understand what he has invoked on the drawing, but all users of the drawing must also understand completely the intent.

The errors appear on drawings today are the same as they always have been, except now CAD and modeling have added new dimensions of problems. It's a little like constipation and diarrhea. The old days were like constipation, really hard and slow, but CAD is like diarrhea, fast and everywhere.

In the usage of GD&T, the primary mistake that always persists is dimensioning from a centerline of a feature and listing of processing and manufacturing information. On a drawing, dimensioning from a centerline is so easy because the crosshair provides a horizontal and/or vertical orientation. In reality, this crosshair doesn't exist. It is normally an axis. This same problem has existed since the beginning of GD&T.

The engineering drawing is used by many people, and it is used primarily as an *inspection document* and a *legal document*. The drawing forms an agreement between the procuring activity (design activity) and a vendor, manufacturer, or subcontractor. The drawing completely defines the end-item requirements of the item. In the endeavor to completely define an item, occasional errors or ambiguities occur that obscure the intended interpretation. These mistakes in interpretation can lead to rejected parts and legal complications. If the rules of ANSI/ASME Y14.5 are followed, then everyone will have the same interpretation. With proper application of simple GD&T and end-item documentation, most of these misinterpretations can be avoided.

The main focus of GD&T is to remove any areas of misinterpretation and to allow looser tolerances. That sounds simple, but the opposite has occurred. ANSI/ASME Y14.5 has many extreme examples and definitions of usages that most of us will never apply. It's a shame that so much of ANSI/ASME Y14.5 is devoted to conditions that are very seldom used. The best thing to do is to keep it simple. Normally, positional tolerances, profile, flatness, perpendicularity, runout, and coplanarity are enough to define an item. An excessive number of datums needs to be avoided, along with the inappropriate use of modifiers. The simpler it is, the less prone it will be to misinterpretation. If you don't understand all the presented data and tolerance zones, then it is very likely others will not understand, either. Always understand all tolerance zones that you have shown, and ensure that all data contains acceptance and rejection criteria.

Always present information in the affirmative manner. Remember that "shall" indicates something mandatory, and one cannot inspect to data presented with "may," "should," or "optional." When information is presented with a "may," "should," or "optional," then it may as well not even appear on the drawing.

All information presented on a drawing shall be capable of being inspected. Never present data that may be vague, ambiguous, or containing a methodology. Review what you listed on the drawing, and ensure that it all contains inspection criteria. There is always the desire is to control the method of manufacture or assembly, but this usually comes with adverse affects. When one introduces a method, then the end result and the method become requirements for inspection. Much of this data is more appropriately shown on in-house processes. A manufacturing process is much easier to change than every drawing where the data is applicable.

With attention to the following outlined areas, many errors will be eliminated. The following is an example of the most common problems appearing on drawings.

THE MOST COMMON ERROR: DIMENSIONS ORIGINATING FROM THE CENTER OF A FEATURE

1. Center of a Feature

A center of a cylindrical object is an axis, not a crosshair. The presentation on a drawing is very misleading. Information originating from an axis has no real orientation unless another feature is utilized (datum) to establish angular orientation (clocking). *Mentally remove the crosshair and visualize an axis.* Another method is, instead of drawing the crosshair as $+$ is to draw it rotated 45° as $\times$, then dimension from the center of the crosshair as before:

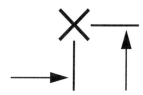

2. Dimensions from a Center

It is very probable that all dimensions originating from a centerline will be incorrect if GD&T isn't used. Use a datum to clarify the origin of the centerline. Normally, the centerline is undefined, or there is no angular orientation (clocking).

3. Identification of Feature

Identify by using GD&T (a datum) to define the centerline from which dimensions are originating. In the feature control frame, specify properly the type of tolerance and shape of the tolerance zone.

4. *Angular Orientation*

Usage of a tertiary datum in the feature control frame is probably required to establish angular orientation (clocking) of the feature when the orientation of features is necessary.

5. *Feature Control Frame*

Use the modifier MMC (i.e., maximum material condition) when applicable and when the additional tolerance is desired. Secondary and tertiary datum callouts list MMC only when that *additional bonus tolerance is actually desired*. If no modifier is specified then, by omission, RFS (regardless of feature size) applies. Do not use the modifier MMC as a default, and fully understand what you have imposed on the drawing.

OTHER FREQUENT ERRORS

6. *Tolerance Zone*

Always understand the shape and location of a tolerance zone for each feature and how an additional tolerance (MMC or LMC [least material condition]) affects it.

All features require both a size and a locational tolerance. Use coordinate dimensioning (±) where possible and GD&T where coordinate dimensioning is not sufficiently accurate or the intent is ambiguous.

7. *Datums*

Always select functional datums. Never use a datum where the feature is inaccessible or of insufficient size for contact. When the datum cannot be contacted because of size or otherwise being inaccessible, then the datum is useless. Care should be taken not to create an excessive number of datums. Normally, three or four are sufficient. Frequently, corner or fillet radii can minimize the actual contactable surface of a datum surface. A datum surface must be accessible. *Example:* Frequently, a mounting hole where a pin is installed is utilized as a datum. If a pin is in the hole, then the hole is not accessible. A better application would be to use an exposed feature on the pin.

Castings

Select datums such that the feature will not be machined and removed at the next level. On the machining drawing of the casting then locate the primary machining datums from the cast datums. If datum points are utilized, this will restrict your application of datums. If datum B is two established points and datum C is one established point, then they should not be used as a primary datum, and datum C should not be used as a primary or secondary datum.

Modifiers—MMC, LMC, or RFS

Do not use MMC as a default. Commonly, MMC is used without any understanding of the consequences. Never use a modifier unless you completely understand how this will affect your tolerance zone and only if you want this effect. Frequently, actual locational tolerance zones can double or triple with inappropriate use of MMC. Overuse of MMC or LMC, without a full understanding, will only complicate the interpretations and compromise your design. It can be very embarrassing if you are asked to explain the exact interpretation to a manufacturer or inspector, and cite the appropriate reference in ASME Y14.5 of your application, but you can't find it or you don't know the interpretation. Remember, this data affects acceptance and/or rejection of parts.

8. Feature Control Frames

If no modifier is shown (MMC or LMC), regardless of feature size (RFS) applies. There is no longer a symbol for RFS, and it applies by omission.

Did you omit the symbol intentionally or was it just an oversight?

9. Tolerance Zone

At the feature control frame for positional tolerance,

if the diameter symbol

is not shown, then the tolerance zone is not round.

10. Basic Dimension

Confusion occurs in the interpretation of a basic dimension. It only describes the theoretical exact size, location, profile or location of a feature or datum target. By itself, it does nothing and it needs to be accompanied with a feature control frame that identifies the accompanying tolerance.

11. End Item

Drawings shall only define end-item requirements. The end result is shown, not the method of obtaining the result. When a method is expressed, then it becomes an inspection requirement for verification, and acceptance and rejection criteria are required. An engineering drawing is a legal document and an inspection document, not a fabrication drawing.

Drawings should not list processes, manufacturing data, quality assurance, or environmental information except in those cases when this information is essential to the definition of the engineering requirements. Whether drawings are prepared for Government contracts or commercial programs, it is a good business practice *not* to provide this information.

When unnecessary information, manufacturing, quality assurance, or processing data is listed on a drawing, or a drawing is structured to mimic assembly or inspection sequences, it imposes a burden on other users of the document (customers or other manufacturers), as it does not allow them the freedom to utilize their normal fabrication, inspection, or procurement and/or subcontractor operations without revision to the drawing. All information that is listed on a drawing becomes a mandatory requirement and requires verification of compliance.

Drawings are prepared to support competitive reprocurement and maintenance for items substantially identical to and interchangeable with the original items. The drawing shall provide the design disclosure information necessary to enable and maintain quality control of items so that the resulting physical and performance characteristics duplicate those of the original design. The drawing shall provide the necessary design, manufacturing, and quality assurance requirements information necessary to enable the reprocurement or manufacture of an interchangeable item that duplicates the physical and performance characteristics of the original product, without additional design engineering effort or recourse to the original design activity.

Drawings should only disclose details of unique processes, not published or generally available to industry, when essential to design and manufacture.

In some cases, processing information may be shown, but that data should be prefixed with an appropriate note such as *"nonmandatory, manufacturing information."* It is imperative to remember that all data on a drawing requires verification of compliance. Information such as *drill and ream, clean, grind, broach, #10 drill,* and *polish,* when not prefixed with this note, becomes mandatory. Evaluate the need to disclose processing information, as it could become part rejection criteria.

Specifying any unnecessary processing, quality assurance, and manufacturing data on drawings should be avoided except when it is essential to describe the item. The following are some of the major areas that may be affected by the disclosure of this information.

a. Inadvertent release of information to sources (competitors) outside of the company, pertaining to manufacturing processes and sequences of assembly. This information should be released only when it is essential to the design or assembly. This information should appear in manufacturing processes, which would be for internal purposes only and not available to competitors.

b. Inspection of an item becomes difficult when a process or a sequence is invoked instead of verifiable end-item requirements. End-item requirements, when specified properly, have measurable parameters that may be verified. However, when a process or sequence is specified, the inspection or verification is difficult, as there normally are no measurable parameters specified. Usable parts may be rejected for failure to comply with a method or a process, and the recovery or rework becomes ambiguous.

c. The listing of quality assurance or inspection information criteria, when not contractually required, places a burden on any other users of the drawing and invokes a mandatory requirement on the current user that may not be deviated from without revision to the drawing. These requirements are normally only in-house procedural requirements and are best placed on in-house documentation such as inspection plans, purchase orders, and manufacturing data sheets. This will allow changes to be made without costly changes to the drawings and will not disclose to other users of the drawings your internal functions and methods of inspection.

12. Acceptance/Rejection Criteria

Evaluate all data on the drawing, especially the notes. Does all data contain inspection *acceptance and rejection* criteria?

13. Using a Profile Tolerance vs. Positional Tolerance

This is done to define curved or irregular surfaces. In many cases, a profile tolerance is desirable in lieu of a positional tolerance (especially in locating radii), and each case should be evaluated to verify which is more functional. *Positional tolerance* locates the center of the feature, while *profile* locates the surface.

14. Is Every Dimension Shown?

The most important item is to verify that all dimensions are present with a tolerance and that they are presented in such a manner as to be subject to only one interpretation.

15. Fillets and Rounds

In models and on the drawings, most features are shown as sharp, with a note defining that all fillets and rounds shall be to X.XX (a specified radius). This is very misleading, as many small features, when these fillet or corner radii are applied, will be nonexistent or a round. Extreme care needs to be taken when specifying fillets and rounds. Visualize the effect of fillets and rounds on features shown square in this part (Fig. 2.1).

16. Stock Thickness

Stock thicknesses should never be listed. A stock thickness is the thickness that the part begins with, not the finished thickness. It is recommended that the material thickness be shown with an appropriate tolerance in a view or note.

17. Coatings and Platings

Clarify if tolerances apply *"after platings and prior to coatings."* A coating is normally a primer or paint. Also, these coatings are known as *organic coatings* if the coating material contains carbon atoms.

18. Dimensional Refinement

Dimensions can be used as a refinement, such as flatness, parallelism, roundness, and so forth. Ensure that these dimensions actually are a refinement to the size tolerance and that the degree of refinement is sufficient for the cost and is a necessity of the design.

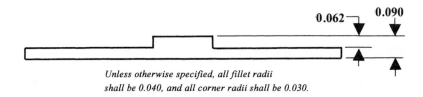

0.062 ⌐   0.090

Unless otherwise specified, all fillet radii
shall be 0.040, and all corner radii shall be 0.030.

Figure 2.1

19. *Nominal*

Avoid usage of this term; there is no definition of the word *nominal*. Although in engineering it is commonly used as a mid range to tolerances, it is not a defined word and is subject to interpretation.

20. *Required vs. Places*

These have different meanings. When stating "required," the term means that, at that particular point of application, the specified quantity is required. For example, "washer, two required" means that, at that particular place, there are "two" washers required (or installed). "Two places" means that "one" washer is used in two different locations. Most companies no longer use the word "places"; they just use the letter "X." Instead of stating "two places," they use "2X."

21. *Drawings Only Depict Perfect Parts*

A drawing defines a perfect part, not a broken, dirty, or out-of-tolerance one, and the information should define it accordingly. Deviations should be shown in manufacturing or inspection processes, not on the drawing. Examples include:

- Grease and dirt shall be removed and part solvent cleaned

- Grind to achieve surface texture

- Touch up paint

These are processes, and they assume that the part is contaminated with grease or dirt, that the paint is chipped, or that the method to achieve the end result must be specified. This requirement cannot be met if the part is not contaminated or the part is not ground. Specified in this manner, the method requires verification in addition to the end result.

22. Revisions/Drawing Changes

Simplify your revisions. All CAD drawings are redrawn whenever they are changed. With appropriately written change notices (ECN/NOR/etc.), the incorporations can be simplified. The days of change balloons and lengthily written revision columns went away 40 years ago. There is no longer a need to write a complete description in the revision block. When contractually permitted, state only the change authority, such as *Revision: A, ECN 12345.*

23. What Requirements Does Your Drawing Have to Meet?

All drawings have to meet some requirements. Don't just assume that the drawing requirements are the same as the requirements of the last program you were on. *Check it out.* As a minimum, there will be internal requirements (i.e., company requirements). Frequently, requirements of programs will vary, even within the same company. In all probability, your drawings will be sent outside of your company to buy or fabricate parts, so it is imperative that they provide for uniform interpretation. The drawing is the book cover of your company.

If the drawing is made in accordance with a Government contract, find the Contract Data Requirements List (CDRL), the appropriate Data Item Description (DID), and any other appropriate information. This will exactly define your requirements. Frequently, the company will extract these requirements and publish them so that they are available to all. Normally, they appear in a Statement of Work (SOW). Regardless of the requirements (company or Government), their intended use is always the same, and you want to provide a universal interpretation. Basically, drawing requirements are all the same, and they stem from ASME documents for their guidelines. (Reference ASME Y14.24, ASME Y14.100, and other sub-tier documents.) The customer (Government) has to prepare the appropriate documentation to specify the Technical Data Package (TDP) requirements for that particular contract. He may want delivery of the drawings, no usage of vendor part numbers, his drawing formats, various associated lists, and so on. Always verify what the customer expects and what your company has agreed to supply.

24. Weight

Often, people feel the need to list the weight on a drawing. This can be a very expensive item and impossible to maintain properly. I recommend that it be placed on the drawing only if required contractually. When not contractually required, list weight as a "reference item" and show only as a "maximum." Weight is really not a problem except temporarily. Once the weight requirements are met, it is no longer important. If not contractually required, no value is added to the design to list the weight. Weight can be derived from the CAD model.

The problem with weight is that it is usually accurate only until the first design change. When a weight is changed, the drawing above (next higher assembly) is also affected. Then, if that assembly is affected, the next assembly also is affected, and so forth until reaching the very top level. We know that management will never sponsor revising every drawing in that particular chain every time a simple design change is made.

When weight is not listed as a reference item, then it becomes an inspectable item, even if it is listed only as a "maximum weight." The problem with the weight being listed as a maximum is that it was probably calculated at nominal values and without finishes, fillets, and rounds. Unless there is a dire need or a contractual requirement, do not list the weight, as it becomes an inspectable item, which can result in rejection of parts. If a part is rejected, rework of a good part becomes nearly impossible—and it was probably just an error in the initial weight calculation.

25. *Effect of Plating or Coatings*

When a part is plated or coated, the drawing shall specify whether the dimension are before or after plating or coatings. Example:

Dimensional limits apply after plating and prior to organic coating.

26. *Undisclosed Test Fixtures and Information*

Avoid reference to undisclosed test fixtures and digital data (software) information. This occurs commonly in drawings, such as test requirements that make reference to a test fixture or digital data without providing the complete part number or drawing where this information is contained. This is always a prime area of concern to any customer and will be the focus of much of his attention. Without disclosure of this information, no one can repeat this test without recourse to the design activity.

3

DIMENSIONING AND TOLERANCING ERRORS

This chapter provides examples of many of the common dimensioning and tolerancing errors on drawings.

Most misapplications of *geometric dimensioning and tolerancing* (GD&T) exist because of improper usage of an *axis,* no angular orientation (no clocking feature), or misapplication and lack of understanding of the implications of using maximum material condition (MMC). The usage of the modifier MMC on *datum features* is commonly used as a default, with no knowledge of the affects. It would be better to use RFS as the default, as there are no additional tolerances incurred, and then there is no effect. The common reason that MMC is used on datum features is that it provides additional bonus tolerances. If you don't know how these additional tolerances will affect your part and dimensioning scheme, then it is best to stay away from them (see Fig. 3.1).

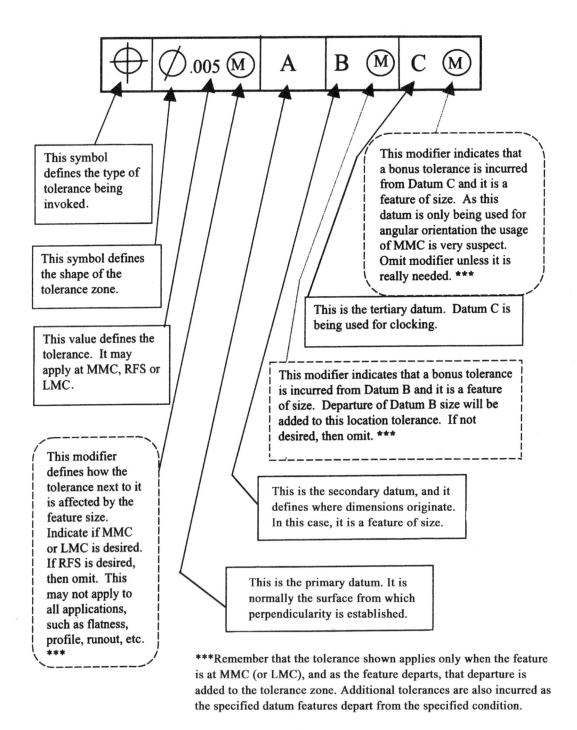

This symbol defines the type of tolerance being invoked.

This symbol defines the shape of the tolerance zone.

This value defines the tolerance. It may apply at MMC, RFS or LMC.

This modifier defines how the tolerance next to it is affected by the feature size. Indicate if MMC or LMC is desired. If RFS is desired, then omit. This may not apply to all applications, such as flatness, profile, runout, etc. ***

This modifier indicates that a bonus tolerance is incurred from Datum C and it is a feature of size. As this datum is only being used for angular orientation the usage of MMC is very suspect. Omit modifier unless it is really needed. ***

This is the tertiary datum. Datum C is being used for clocking.

This modifier indicates that a bonus tolerance is incurred from Datum B and it is a feature of size. Departure of Datum B size will be added to this location tolerance. If not desired, then omit. ***

This is the secondary datum, and it defines where dimensions originate. In this case, it is a feature of size.

This is the primary datum. It is normally the surface from which perpendicularity is established.

***Remember that the tolerance shown applies only when the feature is at MMC (or LMC), and as the feature departs, that departure is added to the tolerance zone. Additional tolerances are also incurred as the specified datum features depart from the specified condition.

Figure 3.1   Feature Control Frame

The center of a round feature is not a crosshair, *it is an axis.* Round features are commonly used as datums. Centerlines (crosshairs) are always shown, but it is the axis that is the datum (see Fig. 3.2).

Use GD&T to define the angular orientation of the holes (when required) and which feature the holes are located from. An angular dimension from the center of the diameter to the holes is unimportant. The angular dimension of hole to hole is normally more critical (Fig. 3.3).

Use GD&T to define the internal features with respect to each other. As stated and shown, there is no angular orientation (Figs. 3.4 and 3.5).

GD&T is required to establish angular orientation of holes and to define from which diameter (feature) the holes are located. In this case, it would be beneficial to utilize a hole as the tertiary datum, for angular orientation (Fig. 3.6).

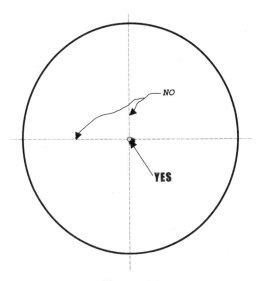

Figure 3.2

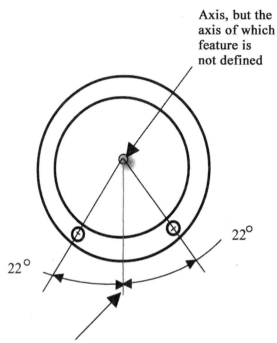

Axis, but the axis of which feature is not defined

22°   22°

To define an angle originating from here (undefined centerline) is rather meaningless. In actuality, it can be any angle. The angle between the two holes is what is critical, and in this case no angular orientation is necessary.

Figure 3.3

Tolerance noncumulative (or nonaccumulative) is an improper method but commonly used to specify hole or feature locations. There is no definition or interpretation of this method in existence. Use positional tolerances in lieu of tolerance noncumulative (Fig. 3.7).

Clearly define from where dimensions are originating. Just a centerline is insufficient unless the feature is clarified. Any dimensions originating from this centerline are originating from an undefined centerline. The centerline can be derived from five dif-

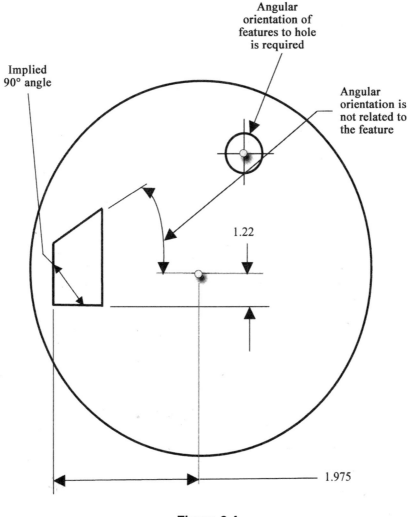

Angular orientation of features to hole is required

Angular orientation is not related to the feature

Implied 90° angle

1.22

1.975

Figure 3.4

ferent features. Use of appropriate datums will define the appropriate datum axis, and a tertiary datum will also provide angular orientation where necessary (Fig. 3.8).

Frequently, an attempt to create a compound datum is performed as shown in Fig. 3.9. It can't be done. Datum holes B and C are an axis, not a crosshair. The axis runs in a different direction. Also, the use of MMC on the datum holes is very suspect.

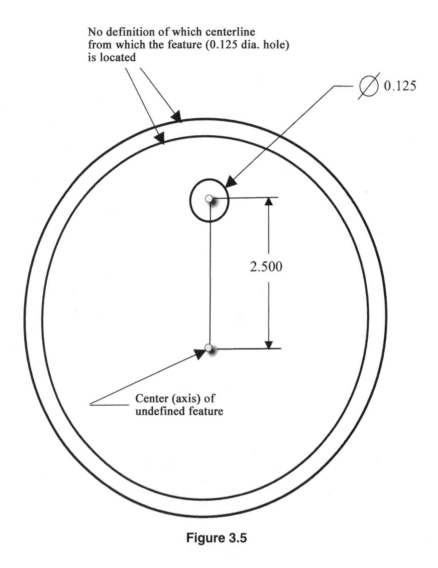

Figure 3.5

Figure 3.10 shows some frequently seen errors: edges not located, slot not located, and angular orientation of hole omitted. As shown the following areas in the figure:

#1: One of these surfaces is not located. A coplanarity dimension will locate these surfaces with respect to each other. Also, it is not defined which surface or surfaces are Datum B.

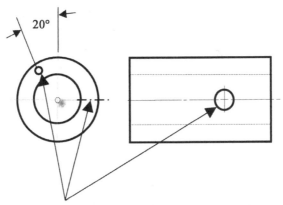

Note: There is no angular orientation of these holes to one another. Perpendicularity of holes also is not defined, both to surfaces and to diameters.

Figure 3.6

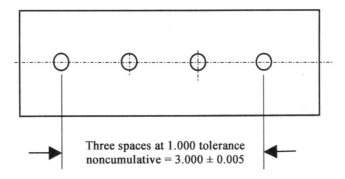

Three spaces at 1.000 tolerance
noncumulative = 3.000 ± 0.005

Figure 3.7

#2: This hole is lacking angular orientation. The hole is perpendicular, located from axis of D, but it has no clocking (angular orientation).

#3: This interpretation of the location of the slot is very complex because of omitted information. It appears that center of slot is located from either the center of Datum D or from the basic 0.61 dimension. If located from Datum D, then there

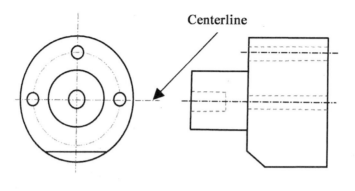

Centerline

The actual centerline of these features
is an axis, not a crosshair.

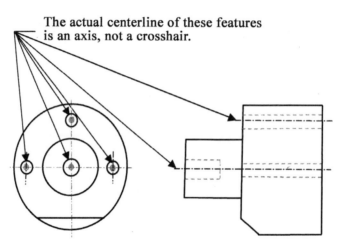

Figure 3.8

is no angular orientation, and positional tolerance is omitted. Also, if it is

located from basic 0.61, GD&T is required. Profile or a positional tolerance

would clarify the intent.

A very common example of dimensions originating from a centerline is shown in

Fig. 3.11. The centerline from which the holes are being located is undefined and

has no meaning. The holes also have no angular orientation with respect to one

another.

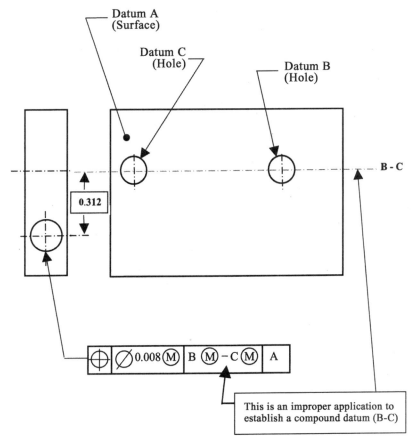

Figure 3.9

Dimensions to or from an Existing Object

Frequently, a feature is stated as being located from the center of an object (such as a pin, screw, etc.), or the part is used as a datum. In Fig. 3.12, it is the center of a pin (or possibly the installation hole). The problem is that it needs to be defined exactly which feature of the pin—such as the outside diameter of the head, the center of the hole, or the small diameter of the pin—is to be used to establish the datum. In all probability, whichever feature is used, RFS should apply. Also, it is important that the datum (fea-

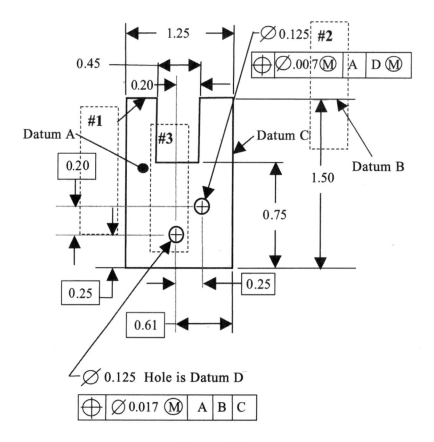

Figure 3.10

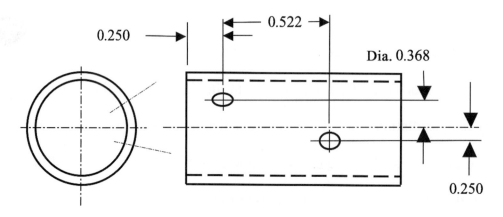

Figure 3.11

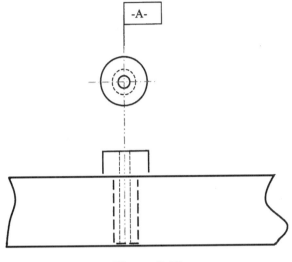

Figure 3.12

ture) be accessible. If the hole in which the pin is installed or the small diameter of the pin is the datum, then it is not accessible.

Other things to consider include flatness/straightness (Fig. 3.13), parallelism (Fig. 3.14), and coplanarity (Fig. 3.15). A commonly missed item is surfaces having no location to each other. Coplanarity (same symbol as profile, but no datum references) is the method of specifying the surfaces' relationship to each other. Point to the extension line between surfaces, use the symbol (same as profile) and the desired tolerances within which the surfaces have to lie. There is no reference to datums. Also, state the number surfaces.

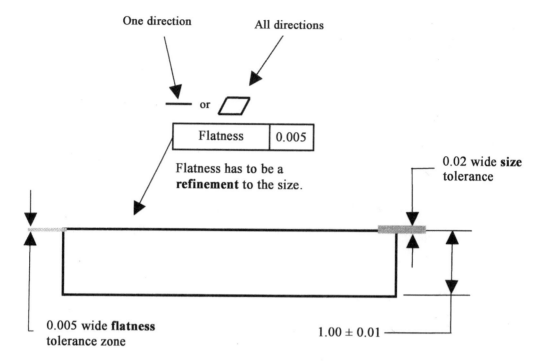

Figure 3.13

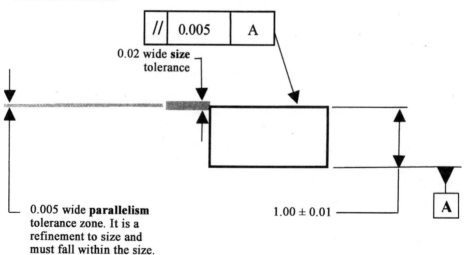

Figure 3.14

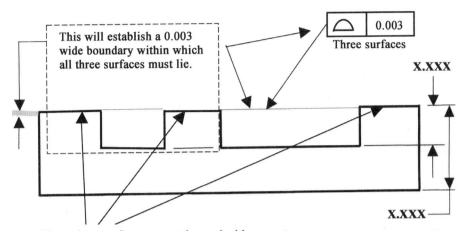

This will establish a 0.003 wide boundary within which all three surfaces must lie.

0.003
Three surfaces

X.XXX

X.XXX

These three surfaces are not located with respect to one another. Use coplanarity symbol to control these three features with respect to one another.

Figure 3.15

4

DRAWING TYPES: SELECT THE PROPER DRAWING FOR YOUR ACTION

There are many different types of drawings, and each one serves a function in the engineering world. These drawings are used to fabricate parts; inspect, test, assemble, and purchase parts; and provide information (such as schematics, installation, interface, interconnect, etc.). Each of these drawing types requires different documentation for the specific item being depicted. The requirement for the fabrication of a casting differs from those requirements for a printed wiring board or a sheet metal part. The documentation requirements of all items originate from ASME Y14.24, ASME Y14.100, and other sub-tier documents that they invoke.

The three most common drawings are *detail drawings, assembly drawings,* and *vendor item drawings*. The most controversial drawing type is the vendor item drawing, as many features are undefined and much of the information is not available.

All the intricate detail requirements of each drawing type will not be completely defined herein, as the requirements are adequately defined within the appropriate governing documents. I will be only stating the appropriate usage of each item, along with some of the basic requirements, and including any pertinent data and the most common mistakes. Always completely review the requirements of ASME Y14.24, ASME Y14.100, and your contract for the type of drawing that you are preparing.

DRAWING STATEMENTS

All Drawings (When Specified in the Contract)

There are many types of drawing statements that may be required on your drawings and parts list. Verify your company requirements and determine whether there are any contractual requirements. Some of the many drawing statements/identifications that may be required are as follows:

- Company Proprietary Statements

- Limited Rights Legends

- Distribution Statements

- Export Control Notices

- Destruction Notices

- Government Purpose Rights

- Critical Safety Items

- Electrostatic-Sensitive Devices

- Explosive Items

These are seemingly meaningless items. However, when omitted by you or your sub-contractors, it can cost enormous amounts of money to correct their omission. I have seen expenditures of millions of dollars to fix a simple omission of these statements.

Additionally, always verify the exact wording of these required statements by reviewing your engineering procedures or contractual requirements.

MONO-DETAIL OR MULTI-DETAIL DRAWING

The delineation of parts may be accomplished by many different methods. Each drawing type is unique and contains its own particular requirements, and when the drawing is prepared as a multi-detail drawing, it must be permitted by the contract.

Occasionally, drawing types or categories are combined on the same drawing. While this may be permissible (verify with your contract), the decision to combine drawing types should be made cautiously. Significant benefits should outweigh such potential disadvantages as:

- Increasing the complexity, which may diminish the clarity and usefulness.

- Accelerated change activity of the combined drawing, which may increase the need to update associated records, material control data, manufacturing planning, and so on.

Mono-Detail Drawings

The use of mono-detail drawings will be specified on the contract or in your engineering procedures. A mono-detail drawing depicts a single part per drawing. Neither the depiction of details on assemblies and inseparable assemblies nor tabulation of parts (detail parts or assemblies) is permitted. A mono-detail detail drawing shall comply with the applicable requirements of its drawing category. When mono-detail drawings have been selected by the contract, it is recommended that the contract be tailored to allow for certain conditions such as inseparable assemblies, various configurations of assemblies, and tabulated drawings.

Multi-Detail Drawings

The use of multi-detail drawing must be permitted by the contract or your engineering procedures. A multi-detail drawing depicts two or more uniquely identified (each part shall have a part identifying number) parts in separate views or in separate sets of views on the same drawing. It is prepared for parts that are usually related to one another. The use of multi-detail drawings should be made cautiously. The same revision status applies to all details on a multi-detail drawing; therefore, a change to one detail of the drawing may affect the associated records of all other details (material control data, manufacturing planning, and so forth). Significant benefits should outweigh this potential disadvantage, as well as such others as diminished clarity and usefulness resulting from increased drawing complexity. See ASME Y14.24 for complete requirements.

DETAIL DRAWING

Part Fabrication

This drawing *completely defines the end product* requirements of a part. It defines configuration, dimensions, tolerances, materials, finishes, marking, surface texture (if necessary), and any mandatory processes. It is important to ensure that all tolerances are applied correctly and that all data contains complete acceptance/rejection criteria. If any dimensions originate/terminate from a centerline, ensure that they are clearly identified with the use of geometric dimensioning and tolerancing. All dimensional requirements are to be specified such that there can be only one interpretation. Review notes to ascertain that all requirements contain acceptance/rejection criteria and that their intent in unambiguous. See ASME Y14.24 for complete requirements.

MATCHED SET DRAWING

For Matched Parts

A matched set drawing is a special application drawing that delineates items that are matched and for which replacement as a matched set is essential. Matched parts are those parts, such as special application parts, that are mechanically or electrically matched, or otherwise mated, and for which replacement as a matched set or pair is imperative. A common depiction is details that are precision machined with other parts at an assembly level, and replacement of the individual parts would not be functional.

A matched set drawing is prepared when the required dimensions, tolerances, or other characteristics of items can only be specified in terms of the matched relationship. This includes items that are interchangeable only as a set because of special requirements for machining, electrical characteristics, performance, and so on. Under such conditions, a matched set drawing defines the matching relationship. Individual parts of the set may be delineated by the matched set drawing or by other drawings.

The matched set drawing shall include, as applicable:

a. The physical or functional mating characteristics of the matched items (set).

b. A unique identifier (part identifying number) assigned to each of the parts and to the matched set. In addition to the part identification marking of the set, the matched parts shall be marked with the word "set" next to the part identifying number.

c. The statement "furnish only as a matched set" or similar note shall be included on the drawing.

d. The drawing shall comply with all the other requirements for the drawing category.

See ASME Y14.24 for complete requirements.

ASSEMBLY DRAWING

Assembled Parts

There are two different types in this category: a separable assembly and an inseparable assembly. See ASME Y14.24 for all requirements. A detail part that contains threaded inserts, riveted parts, and welded parts/members is normally construed as an inseparable assembly.

The intent of an assembly drawing is to define the item as an end product. The data shall be presented in such a manner that the requirements are not subject to more than one interpretation. Assembly drawings should not be structured to mimic assembly or inspection sequences. Drawings should depict logical levels of assembly or disassembly, at testable levels, as a functional item or a deliverable item.

Assembly drawings should not contain noncritical manufacturing processes or be structured to mimic assembly or inspection sequences, as previously discussed, as increased costs may be realized such as the following:

a. The total quantity of drawings requiring preparation, maintenance, and delivery to the customer will increase.

b. Maintenance costs of the drawing increase as a result of of additional engineering changes being written against the documentation so as to revise processing data or manufacturing information as techniques change or as changes to equipment availability occur. When the drawings are structured logically and manufacturing process information is omitted, then manufacturing may revise its own processes without the costly burden of revising the drawing on each occurrence.

c. *Spares levels.* When assemblies are not created logically, additional maintenance, repair manuals, and testing data are required to support these extra intermediate levels.

d. *Stocking levels.* Intermediate assembly levels will be required at each of these levels to support the sparing requirements, which creates additional administrative and manpower burdens.

e. *Testing levels.* Tests and equipment are required to support each level, along with appropriate documentation. Test requirement documents, specifications, and technical manuals may be required to support these additional levels.

f. *Reference designations.* Assignment of unnecessary and/or improper reference designations complicates the identification of items and their locations. These designations may become so large that they become illogical and difficult to mark on the equipment and list in technical manuals as well as system and schematic diagrams. See Chapter 5.

Attaching Parts

Attaching parts (bolts, nuts, washers, and so on) required to mount assemblies into their next higher assemblies or on foundations shall be called out on the parts list of the drawing that defines the attachment (usually the next level assembly or installation drawing).

Reference Items

A cross reference to applicable installation drawings, wiring lists, schematic diagrams, test specifications, and associated lists shall be provided as applicable.

Assemblies Containing Electrical Parts

Testing data in the form of a stand-alone test specification or test requirement is normally required on assemblies where *active circuitry* is employed. Testing data is of extreme importance to customers and will become an area of intense inquiry.

Caution should be exercised to restrict testing to actual testing requirements and not to use electrical inspection requirements as testing data. A common occurrence is an "electrical continuity test" to determine if all electrical connections have been made on wiring or circuitry. The actual requirement on the drawing is to make all the connections listed, include the wire or circuitry to be unbroken, and note that the soldering is normally performed to a governing specification. An electrical continuity test is solely an inspection to verify that the assembly has been properly fabricated—it is not a test.

Parts of the Assembly

Locate all parts and materials and any special requirements that are performed or installed at this level.

Electrical Parts of the Assembly

Identify, in the field of the drawing, the reference designation assigned to the respective parts in the parts list (BOM). See Chapter 5 for reference designations.

Quantity of Parts

Quantities in the parts list (BOM) shall agree with the field of the drawing.

Notes

Notes shall be unambiguous and contain complete acceptance/rejection criteria. See ASME Y14.24 for complete requirements.

INSEPARABLE ASSEMBLY

This is the same as an assembly drawing except as noted below.

An inseparable assembly, although documented as an assembly drawing, actually is a part and is identified as a part. The inseparable assembly drawing delineates items (pieces) that are separately fabricated and are permanently jointed together (as in welded, brazed, riveted, sewed, glued, or otherwise processed) to form an integral unit (part) not normally capable of being disassembled for replacement or repair of the individual pieces.

An inseparable assembly drawing may be prepared in lieu of individual mono-detail drawings for inseparable parts, but it must be allowed by the contract, either by the allowance of multi-detail drawings or tailored when mono-detail drawings have been selected as an option in the contract. Always verify within the contract that it has been tailored to allow multi-detail drawings for inseparable assemblies.

The drawing shall disclose all the required data of an as-assembled, end-product item, complying with all the applicable requirements of a detail drawing and assembly drawing (including a parts list or equivalent disclosure of individual parts). Materials, finishes, envelope configuration, complete dimensional characteristics, and so forth shall be provided.

As the item is being defined as an end-item product, frequently many of the individual features may be omitted and left to the ingenuity of the fabrication (such as end joint features of a welded or brazed assembly), with only the end-item product being defined. The detailing of the individual pieces on separate drawings is not normally a very cost-effective method. See ASME Y14.24 for complete requirements.

DETAIL ASSEMBLY

A detail assembly drawing depicts an assembly on which one or more parts are detailed in the assembly view or on separate detail views. Even though this is a widely used drawing, it is a drawing category that no longer is recognized in ASME Y14.24 and falls into the category of a multi-detail drawing. The preparation of a detail assembly drawing in lieu of individual mono-detail drawing must be allowed by the contract (Government contracts only), either by the allowance of multi-detail drawings or tailored when mono-detail drawings has been selected as an option in the contract. It must be tailored to permit the use of multi-detail drawing for a detail assembly.

The detail assembly drawing shall provide complete end-product definition of each individual part and the assembly, as each part may be spared as a replaceable item.

The drawing shall contain all the end-item disclosure information of an assembly drawing and detail drawing. Parts shall contain complete definition of configuration, dimensions, tolerances, materials, finishes, marking (if necessary), surface texture (if necessary), and any mandatory processes. Ensure that all tolerances are applied correctly and that all data contains complete acceptance/rejection criteria.

ALTERED ITEM DRAWING TO ALTER AN EXISTING ITEM

This drawing physically alters an existing item (a usable item) under the control of another design activity. You cannot alter your own items or items under your design control. If this were permitted, you would build a part incorrectly and then perform the modification.

The item's form, fit, function, and performance requirements prior to the alteration must be described in some manner, and you must provide a complete definition of the alteration. This can be accomplished in many ways. As the item existed prior to the alteration, it may be described on the actual altered item drawing, on a separate drawing, or by a vendor part number.

- Always use caution when altering a vendor-supplied part; by virtue of this alteration, any warranty may be compromised.

- The original part number should be removed or obliterated and the part re-identified with the new part identifying number.

- The alteration may be performed by you or by another competent manufacturer, including the original manufacturer or a third party.

See ASME Y14.24 for complete requirements.

PRINTED WIRING BOARD DETAIL DRAWING

A definition of the circuitry is required. Circuitry definition may be in the form of a software drawing, software (digital data), a separate master pattern drawing, or inclusion of a representation of the circuitry, from the printed wiring master drawing, as subsequent sheets of the drawing. The depicted circuitry is the circuitry prior to etching and not that of the actual finished printed wiring board. The preferred method is to furnish software (digital data) that defines the circuitry. The use of a master pattern drawing and definition of the circuitry as subsequent sheets is not preferred. Storage problems and usability are the main issues with anything other than a software definition.

The end-item requirements, materials, and physical shape of the board require definition. This is normally straightforward. The board outline is normally defined using basic dimensions and a profile tolerance, originating from a surface and two datum holes (indexing holes or tooling holes). A board edge should never be used as a datum. The board edge is actually located from the primary datum hole, and the symbol of a "dimension of origin" is frequently used for this clarification.

Holes

The holes shall be located dimensionally or by annular ring requirements. Holes that have no conductive material (unsupported holes) and holes that are in large circuit patterns/pads require dimensional locations, as the annular ring requirement is not sufficient. It may be unrealistic to define the size and location of small via holes (such as those used in printed wiring boards utilizing surface mount components or interconnects) as an inspection requirement, as they will probably be solder filled. It is recommended that these particular holes be defined only by reference dimensions or allowed to be located by annular ring.

Terms

Confusion also lies with many of the terms used in printed wiring board documentation. The following are definitions of some of these key terms that may affect documentation.

Annular Ring   That portion of conductive material completely surrounding a hole.

Printed Wiring Drawing   This was previously known as a *printed wiring master drawing* or *master drawing*. It is the drawing that shows the dimensional limits or grid locations applicable to any or all parts of a printed board (rigid or flexible), including the arrangement of conductive and nonconductive patterns or elements; size, type, and location of holes; and any other information necessary to describe the product to be fabricated.

Artwork Master, Printed Wiring Master Pattern Drawing (Stable Base Artwork) The master pattern drawing is a reproduction of the original artwork or database prepared on a drawing format

Production Master A one-to-one scale pattern used to produce one or more printed boards within the accuracy specified.

Via Hole A plated-through hole used as a through connection, but for which there is no intention to insert a component lead or other reinforcing material.

See ASME Y14.24 and sub-tier ANSI/IPC standards such as ANSI/IPC-325, ANSI/IPC-D-275, ANSI/IPC-D-350, and ANSI/IPC-D-351 for a complete definition of all requirements.

CONTROL DRAWINGS

When You Want to Procure a Part from a Subcontractor or Vendor

When you want to purchase a vendor part, the documentation will appear on one of these following drawings. The most common drawing is a vendor item drawing. This drawing buys a part off the shelf with no restrictions other than what the vendor normally provides.

An alternative to preparing a drawing is to list the vendor and his part number on your parts list. The problem that arises when only a vendor part number is used is that you are restricted to buying only that part, and competitive procurement is eliminated. Additionally, verify that you are contractually permitted to list a vendor part number in lieu of preparing a drawing.

A control drawing normally does not cost much to prepare and, when done properly, is much preferred to listing a vendor part number on a parts list.

VENDOR ITEM DRAWING

To Buy a Part Right off the Shelf

If you have a Government data requirement that requires that you prepare vendor item drawings, this will be the most controversial drawing in your technical data package, because it does not provide a complete item description. You are only defining what the vendor guarantees. This drawing provides an engineering description and acceptance criteria for commercial items or vendor-developed items that are procurable from a specialized segment of industry.

When you want to buy a part from a vendor and it comes right off the shelf, then you prepare a vendor item drawing. This is just like buying it from a store. It is just as they advertise it, with no modifications, selections, or restrictions that they do not advertise or guarantee. This drawing allows procurement to have competitive bids from multiple vendors. Purchasing can buy from anyone who meets the qualifications you have listed on the drawing. List the vendor (two or more if possible), the physical definition, and any other pertinent information from the vendor. Never tighten tolerances beyond what the vendor guarantees or relax any noncritical dimensions and characteristics. The most important item on the drawing is the name of the vendor and his part number. Regardless of what other information is on the drawing, the part number is the item that is procured. The information in the notes and the outline is for inspection purposes only, to verify that the part procured complies with the information depicted on the drawing.

In lieu of a drawing, you could list a vendor part number in the BOM/parts list, but that restricts purchasing to buying only that part from the vendor listed. This action can be taken only when contractually permitted.

Do not list recommend mounting holes on the drawing. Many times, the vendors data sheets contains errors, and there could be slight differences, depending on who supplies the part.

The part identifying number is the vendor's part number, not your part number (although, in many cases, a vendor will use your part number as his part identification). You may desire to have your part identification, in addition to the vendor part number, marked for administrative purposes.

Do not place requirements in addition to those normally provided by the vendor. When restrictions are placed on a part beyond what the vendor guarantees, then it will no longer comply with the parameters defined by his part number. Relax tolerances and requirements that are not critical to your application, but completely define all critical requirements, provided that they are guaranteed by the vendor.

See ASME Y14.24 for complete requirements.

PROCUREMENT CONTROL DRAWING

To Buy a Part in Development and Establish a Part Number

This is something of a mix between a vendor item drawing and an envelope drawing, except that it can establish a part number. It is used for vendor or subcontractor items under development when engineering needs a part number to list in their system. A procurement control drawing may be prepared in lieu of other types control drawings to specify criteria for:

- A purchased item

- The alteration of a purchased item or an item defined by a nationally recognized standard

- The selection of a purchased item or an item defined by a nationally recognized standard

- The development and qualification of a new item

- Item identification

Data should be provided to properly define the item, including performance, envelope dimensions, procurement information, and qualification and acceptance requirements. The drawing should contain all the necessary information to define the requirements of your application.

See ASME Y14.24 for complete requirements.

ENVELOPE DRAWING

For Items Subcontracted for Development

When a part is subcontracted to a vendor or manufacturer, necessary information (such as performance, interfaces, envelope size, and any other pertinent data) is shown on the drawing. This drawing shall provide the basic technical data and performance requirements necessary for development or design selection of an item.

No part number is established at this time. If you attempt to establish a part number prior to development being completed, it will not be known whether to use a detail part number or an assembly, vendor, or source control number.

When development is completed, the drawing shall evolve into formal documentation, such as a control drawing of some type, a detail drawing, or an assembly drawing.

See ASME Y14.24 for complete requirements.

SOURCE CONTROL DRAWING

To Restrict Purchase of a Vendor's Item

This drawing has once-in-a-million employment, so it will not be discussed extensively. This type of drawing defines a unique item that requires prior qualification in advance of procurement and restricts procurement to only the vendor listed.

It provides an engineering description, qualification requirements, and acceptance criteria for commercial items or vendor-developed items procurable from a specialized segment of industry. It provides performance, installation, interchangeability, or other characteristics required for critical applications.

Never prepare a source control drawing and list yourself as a sole source of the item. The requirements are nearly identical to those of a vendor item; however, there may be unique Government approval requirements.

See ASME Y14.100, ASME Y14.24, and your contract to observe any special requirements that may apply.

ASSOCIATED LISTS

Parts List (PL)

Parts lists are usually not a problem, and whatever format you use is normally acceptable. It has been time proven and it works, or it wouldn't be there. If prepared for a Government contract, consult ASME Y 14.34 to ensure compliance. In areas of non-compliance, it is easier to request a waiver than to change your entire system for a single program. If, on your contract, you are required to prepare data lists and/or index lists, review ASME Y14.34 to verify your ability to prepare them properly. They are very tedious and complicated documents to prepare properly. Most people will never come into contact with a data list or index list; these documents are used primarily by your customer.

Data List (DL)

A data list is a very complex item. It will list all engineering drawings, associated lists, specifications, standards, and subordinate data lists pertaining to the item to which the data list applies, along with essential in-house documents necessary to meet the design disclosure requirements—except for in-house documents that are referenced parenthetically. In actuality, a data list is a complete listing of all information appearing on your drawing. The required data includes *document number, revision status,* and *nomenclature.* Optional information is *drawing size* and *number of sheets.*

Index List (IL)

An index list is not normally required in a contract. However, when required, it is a tabulation of data lists and subordinate index lists pertaining to the item to which the index list applies.

DRAWING NOTES

The drawing notes provide pertinent data in word form to complement the delineation or to clarify the requirements of the parameters (acceptance or rejection criteria). They do not contain ambiguous, uninspectable parameters. Controls shall be established for all variables that could affect the accuracy of test rest results. All parameters are to contain a tolerance or a range of values (directly or indirectly) that are inspectable.

Notes shall be short, concise statements, using the simplest words and phrases for conveying the intended meaning. The use of abbreviations should be minimized and confined only to recognized and authorized abbreviations. Sentences should be written in the affirmative manner where the requirement is mandatory. Requirements shall be specified as requirements, not methods or actions. For example,

Proper: Part shall be free of burrs and sharp edges.

Improper: Remove all burrs and sharp edges. (This is an action, not a requirement.)

Commonly, the drawing notes supplement the drawing by providing the following information. This is a list of the types of items specified in the notes. It is not all inclusive; it provides just the most common information.

- Material

- Finish

• Supplemental dimensional data

Surface texture

Fillet and corner radii

Sharp edge and burr requirements

Marking requirements

Mandatory manufacturing information

Material processing requirements such as stress relieve

Cautionary information

Applicable geometric dimensioning and tolerancing (applicable specification and date) applicability

Installation information for rivets, inserts, pins, etc.

ISOMETRIC VIEWS (VALUE QUESTIONABLE)

Isometric views, although not really required, can add some noteworthy information to the drawing. They are relatively free, but their value is questionable. *If not maintained, they will become incorrect as a result of drawing changes.*

Datum Location

The primary advantage is that this can be used to display a quick look at the *location of datums*. This provides quick orientation as to their location, as opposed to sorting through many sheets or views to locate datums.

Occasionally, other bits of key information (markings, parting lines, unique finish requirements, location of parts of an assembly, and so on) are displayed on this view.

The view should be labeled as *"Reference View"* only, and any of the appropriate data displayed on the isometric view shall be shown on the appropriate views of the drawing.

Frequently, the isometric view is utilized in the preparation of technical manuals or other similar information. On an assembly drawing, they provide excellent information that can also be exported into technical manuals. Frequently, exploded isometric views are used in lieu of the traditional orthographic views. Verify that there is no conflict, contractually or internally, before using these views.

5

ELECTRICAL REFERENCE DESIGNATIONS

Electrical items are assigned an identifying letter and number for quick identification and location in equipment. Each type of electrical item has a specific letter assigned for quick identification. For example, a resistor is assigned the letter "R." Then each item is assigned a number identifying it uniquely. When this is all put together, it specifies exactly where an item is located. It is really nothing more than an address with an identifying letter of the occupant. Take, for example, a resistor R1. When the complete prefix is used, "1A2A3A1A1R1" will quickly reveal exactly where this particular component is located. Think of this as your address:

R1	This is you
A1	558
A1	Main Street
A3	Chicago
A2	IL
1 (unit number)	U.S.A.

The marking of reference designations is extremely beneficial for rework and troubleshooting of equipment. Normally, only a partial reference designation is marked, and it is placed adjacent to the item.

In this chapter, I provide only basic reference designation information. For more advanced or specific applications, see the appropriate governing specifications, ASME Y14.44 and IEEE Std 315.

HOW ASSIGNED

The appropriate reference designation is assigned to items *via the parts list* (BOM) except for the top (end) item, which has no higher assembly or parts list. All electronic parts and equipment must have a reference designation, which shall appear on the parts list.

Parts List

Reference designations are assigned by, and appear on, the parts list for all electronic parts of that particular assembly, and they do not carry any prefix. For example, R1 (not A1R1), W1 (not A1W1), A1 (not A1A1). The numbers of the reference designation should be sequential. For example, A1, A2, A3, not A2, A4, A6.

Item Identifiers (Find Numbers) and Reference Designations

In the parts list, electronic items are assigned an item identifier (a "find number," and it normally is a number) and a reference designation. In some systems, the items are listed in the parts list by reference designation in lieu of an item identifier (find number). It is preferred to identify the location of the item in the field of the drawing by the reference designation and/or the find number (balloon) or a combination of both. Use of a reference designation only is especially beneficial in the case of circuit card assemblies, while electronic assemblies and cable assemblies may use a combination. Identification in the field of the drawing may be either in a wiring diagram (when one is included as part of the assembly) or in the item delineation. Using only the reference designation of bulk type items (such as splices, terminals, shield terminations, and such) that are identified in the wiring diagram with a reference designation is very beneficial and much more understandable than placing an item identifier (balloon) in the item delineation.

The only intent of an item identifier (find number) is to identify the location of an item in the assembly. The intent of a reference designation is the same, except it also identifies what type of an item it is and its location in a system in addition to its location at the level of installation.

WHICH ITEMS RECEIVE A REFERENCE DESIGNATION?

Each electronic item/electronic assembly is assigned a reference designation. A reference designation defines the type of item it is and, by use of the *full* reference designation, it will also define exactly where it fits into the equipment.

For example, cable is assigned *W*, resistors *R*, capacitors *C*, repairable electronic assemblies *A*, non-repairable electronic assemblies *U*, and so on. IEEE Std 315 defines the various reference designations approved for use (Table 5.1). This system has been in place since the beginning of time, or maybe five years before that, and it has never changed. This system makes items easily identifiable. It defines what an item is and where it is located in the equipment. A full reference designation is like an address on a map, and it will identify its exact location. Reference designation information is also used to create the mates with information on band markers of cable assemblies, and it is used in interconnect diagrams, test requirements, test specifications, schematics, family trees, and extensively in technical manuals and technical orders.

If you have no contract that invokes IEEE Std 315 or ASME Y14.44, then all you have are your own engineering procedures and/or common sense. See the listing of the most common reference designations. For a complete listing and definitions, see IEEE Std 315. Never create your own reference designation if one already exists, as this only creates confusion. These designations are universally understood and recognized.

REFERENCE DESIGNATION STARTS

Wherever a reference designation starts, it continues upward in the system, *with no omissions*. For example, a circuit card assembly (CCA) assigns reference designations to the various components on it. The assembly above that picks up the CCA and

Table 5.1   Commonly Used Reference Designations (See IEEE Std 315 for Complete Listing)

Component	Reference Designation
Amplifier	AR
Assembly (repairable)[a]	A
Assembly (nonrepairable)[b]	U
Cable assembly	W
Capacitor	C
Connector (receptacle connector)	J
Connector (plug connector)	P
Diode	CR or D
Fan (blower)	B
Filter	FL
Fuse	F
Heater	HR
Inductor	L
Jumper wire	W
Microcircuit	U
Motor	M
Oscillator	Y or G
Resistor, thermal	RT
Splice	E
Switch	S
Terminal board/strip	TB
Test point	TP
Thermistor	RT
Transformer	T
Waveguide	W
Wire	W

[a]See IEEE Std 315, Para. 22.2.4 for definition of repairable when in doubt.
[b]See IEEE Std 315, Para. 22.2.5 for definition of nonrepairable when in doubt.

assigns a reference designation to it. The next assembly that picks up that assembly assigns a reference designation to that assembly, and so forth, until reaching the very top level. Do not assign a reference designation for an item at the item level. *The reference designation is assigned on the parts list at the level at which the item is installed.* For example. a drawing of a capacitor does not assign a reference designation at the detail level. The capacitor may have multiple usages, and the level at which it is installed will determine each reference designation.

UNITS, SETS, AND NOMENCLATURE

Top deliverable end items shall be assigned a unit number or a set number. It depends on the contract (Government), but usually it is a unit number. The contract (if a Government contract) defines whether it is a unit or set. Normally, there is a Unit 1, but there could also be a Unit 2, Unit 3, and so on (Figs. 5.1 and 5.2). At that point, everything trees downward to the lowest level. If it is a set, such as a navigation pod and a targeting pod, and if they are delivered as sets, then it would be Set 1 and Set 2 instead of units. The contract should define if it is to be Sets or Units. In addition, *assigned nomenclature* (not to be confused with drawing title) should be defined, or nomenclature may have to be requested. (See Mil-Std-196 or equivalent.)

Set   A set is a unit or units and necessary assemblies, subassemblies, and parts
    connected or associated together to perform an operational function.

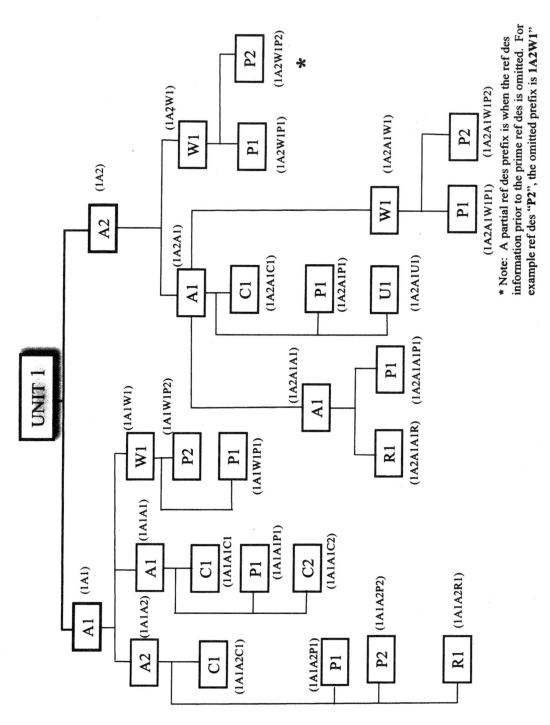

* Note: A partial ref des prefix is when the ref des information prior to the prime ref des is omitted. For example ref des "P2", the omitted prefix is 1A2W1"

Figure 5.1

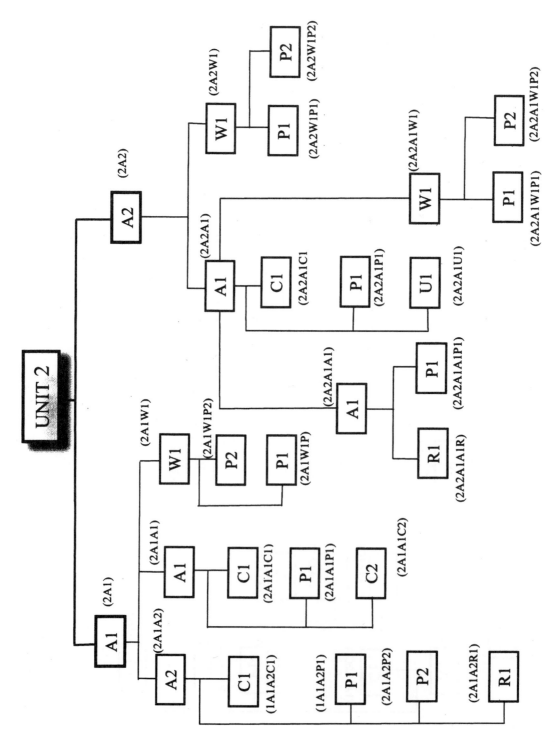

Figure 5.2

Unit   A unit is a self-contained collection of parts and/or assemblies within one package, performing a specific function or group of functions and removable as a single package from an operating system.

ASSIGNING REFERENCE DESIGNATION

Start with the top item and then assign a reference designation to each electronic item that appears on each assembly beneath it. Tree it all the way down until there are no more assemblies with electronic items (see Fig. 5.3).

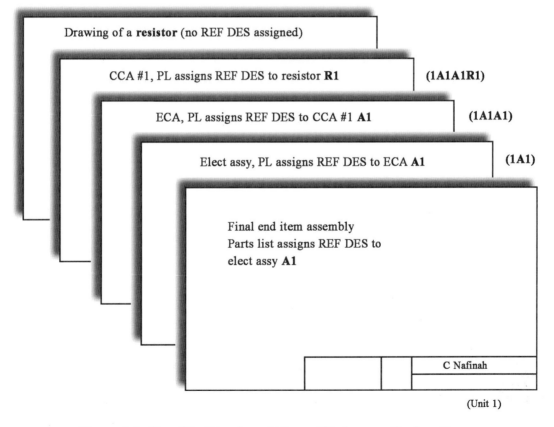

Drawing of a **resistor** (no REF DES assigned)

CCA #1, PL assigns REF DES to resistor **R1** (1A1A1R1)

ECA, PL assigns REF DES to CCA #1 **A1** (1A1A1)

Elect assy, PL assigns REF DES to ECA **A1** (1A1)

Final end item assembly
Parts list assigns REF DES to
elect assy **A1**

C Nafinah

(Unit 1)

Figure 5.3 Simplified Version of Flow of Reference Designations

CABLE ASSEMBLIES

On cable assemblies, identify the "mates with" information. It is recommended to identify only as much of the prefix as is necessary to prevent improper hookups. Instead of "mates 1A1A1W1P1," "W1P1" is probably sufficient. Attempting to synchronize a connector reference designation to the reference designation of the mating connector is futile, and it really simplifies nothing. For example, assignment of a reference designation to a connector such as "P14" because it connects to"J14" is not beneficial. It just raises doubts of what happened to P1 through P13. With adequate "mates with" marking, the problem is easily resolved.

MARKING

Marking of reference designation designations is normally accomplished by placing the marking adjacent to the item and not directly on the item. ASME Y14.44 established the formulation and identification requirements of reference designations for electrical and electronic parts and equipment.

CONNECTOR, WIRING SIDE

Usage of the reference designation "X" denotes wiring side. The prefix "X" is normally used to denote the wiring side or the side where an item plugs in. It is not a *"stand-alone"* reference designation but only the indicator used to identify a location where the item plugs in. For example, connector P1 plugs into XP1. The connector is still referred to as P1, not XP1. XP1 identifies the location where P1 is installed.

PARTIAL REFERENCE DESIGNATIONS

Reference designations, assigned to the parts list and shown on the drawing, are only partial reference designations, as they require all the reference designations above them to be a complete and full reference designation. To clarify this, it is required and common practice to place a note on the drawing such as, "Partial reference designations are shown." For full reference designation, prefix the reference designations with the unit number (or set number) and subassembly designation.

6

INSPECTION

THE INTEGRITY OF YOUR SUPPLIER: WHO INSPECTS AND WHAT IS INSPECTED

Inspection is always a gamble—which features to inspect and which parts are inspected. There really is no requirement to inspect. Inspection is derived as a reflection of confidence in your fabricator. The reality is that most parts will receive only cursory inspections by the procuring activity (design activity). Most inspections will be performed by the fabricator or vendor. The reasons for this vary greatly, but it is primarily dependent on that availability of equipment for inspectors in the field. Many features may go uninspected, as 100 percent inspection is not realistic. Many of the uninspected features are noncritical, but critical features frequently may be overlooked. Incoming (in-house) inspection at the procuring activity has more capability, but in many cases this is just too late, and inspection/rejection causes more delays. In all probability, parts most likely will be reworked in house instead of rejecting them back to the fabricator or vendor. When parts are rejected at the procuring activity, many delays, or even line stoppages, will be incurred.

A drawing normally does not specify any inspection criteria. What is inspected, if anything, is entirely up to the procuring activity and the confidence you have in your fabricator or vendor. You may elect to have full inspection until you have developed sufficient faith in the fabricator/vendor. Any inspection less than 100 percent is strictly a gamble, and you have to feel confident for whatever level of inspection you perform.

It is recommended that *critical features be identified* in some manner and denoted on the drawing or in a separate document (an inspection plan or the like). In this document, denote that specified features shall receive a 100 percent inspection, 75 percent inspection, 50 percent inspection, and so forth. When this data is specified in a separate inspection plan, it can reduces the amount of change activity to the drawing. When the data is placed on the drawing, an engineering change is required to make any revisions. It is much easier to change one inspection procedure than 100 drawings.

7

DELIVERY OF YOUR DRAWINGS (TECHNICAL DATA PACKAGE)

In this chapter, I will explain how delivery of the drawings that you prepared impacts the delivery of the technical data package to the customer. If you are not on a Government program that requires delivery, then this chapter is not significant to you. In past years, almost all Government contracts required delivery, and now this requirement has been reduced to minimal delivery requirements. There is always a movement to revert back to the days of the past delivery requirements.

There are many variables in the delivery of the technical data package (TDP) and in the drawing preparation requirements. Always be aware of what your requirements are, either internal or customer.

This is not the first chapter of the book; however, it should be treated as a "read me first" type of subject. Your standard drawing practices and the following topics should have been discussed at the beginning of the contract or at the time of the contract award to avoid problems at the time of drawing delivery. Clarifying your practices with the customer at the beginning can minimize the most common discrepancies and mis-understandings that may occur at the time of delivery of the technical data package. If these items are resolved or clarified in the beginning with the customer, huge savings may result, and transmittal of the technical data package will be facilitated. Careful scrutiny of the contractual requirements and your ability to comply to these require-ments at the beginning of the contract allows you to plan accordingly in your docu-mentation and subcontracting efforts. Never be in the situation at the end of a program of finding that your practices or subcontractors are in violation of the contract.

The following are the most significant problems that require resolution or close attention during the preparation of the engineering documentation so as to minimize their impact at the time of delivery.

VENDOR PRINTS

If you used vendor part numbers anywhere in your documentation, you may be expected to provide a copy of those drawings—or is your customer expected to obtain them? Will they comply with the legibility and documentation requirements desired/ required by the customer? Are these documents expected to comply with the same level of drawing requirements as specified in the contract, or will they be acceptable in the as received status? If you know the answer to these questions at the time of the contract award, much grief and expense can be avoided. Attempting to rectify erroneous interpretations of what the customer desires at the end of the program may result in huge manpower and money expenditures.

MISSING DRAWINGS

This always sounds like an insignificant item, and it probably is nothing more than a logistical problem. However, in some cases it may be a deeper-rooted problem than it appears. Confusion between specifications and drawings always abounds, which normally is a result of documentation practices.

Design activities always want to make certain drawings look like specifications, because it looks professional, and they attempt to process actual specifications like drawings. True specifications are prepared per Mil-Std-961 and are normally delivered separately and as specified in the Contract Data Requirements List (CDRL). If it is on a drawing format, it is not a specification, it is a drawing! If it is looks like a book and is not on a drawing format, then it is a specification. This delivery of specifications is normally by either hardbound paper delivery or in the form of digital data. The problems begin when these specifications are listed on drawings or appear on data lists under the heading of drawings. This confuses the customer, and then he expects to see the specifications delivered along with the rest of the technical data package and in the same delivered form.

UNDISCLOSED DATA

Undisclosed data is another area to be avoided. The reference to unique, undisclosed information (such as test fixtures, test circuits, digital data, procedures, and so forth) does not fulfill the requirements of the contract and, when discovered at the termination of the contract, may result in delayed payment by the customer until this information is provided.

DELIVERY OF DIGITAL DATA AND MODELS

Ensure that delivery of models and digital data is done per the contract. Also, ascertain whether the customer can receive, store, and retrieve this data as necessary.

Delivery of the technical data package is usually the final phase of your engineering efforts, and normally that is when you get paid. Additional deliveries will probably be required to provide the balance of drawings, or as changes occur to the data package on a prearranged scheduled basis, such as every 90 days.

Many problems can occur, but careful scrutiny of missing drawings, undisclosed data, and vendor prints can eliminate the majority of the problems.

8

THE CHECKER: WHAT HE WILL LOOK FOR

The checker who will review your drawing has a different mindset from yours. You are optimistic, your drawing is right and there is not a mistake on it, it is perfect. However, the checker knows your drawing is not perfect but that there is something right on it and he just has to find it. He will sift through your drawing notes, the delineation, geometric dimensioning and tolerancing, and the parts list. All data is reviewed for correctness. He has to evaluate any errors and their significance. If petty and relatively unimportant, he may discard them. He has a mental checklist that he uses, and it is different for every drawing type. A competent checker can find 75 percent of your errors in less than a half an hour, and if you know what he looks for, then you will be many steps ahead.

Checking is always a very difficult task and takes a very special individual. He must overcome his desire to redesign as well as being flexible for the various types of checking needs. He may be asked to perform a full-up check, design analysis, tolerance stack up, or verify that all contractual requirements are met for a production level contract. He may be expected to perform only a quick check for specific requirements for a very limited production, or he may be looking at just a quick drawing created to obtain quotes. He must be willing to delve into intricate details until he is completely satisfied that all errors have been discovered.

The more cooperative the designer is in providing information, the better the end item will be. The better the end item, then the better you will look. No one may remember that you did a good job, but they will definitely remember your name if you did a bad job.

The first item that a checker will need to determine is what the proper drawing requirements are. Is it for a contract, or is it to in-house requirements? He will then start a systematic check. His mental checklist may be as follows (it will vary for each particular type of drawing, and this is only a partial listing):

❑ Check drawing title, sheet numbering, revision information, title block information, including contract number (if applicable), appropriate notations ("see separate parts list," etc.) and drawing statements (limited rights, export notices, etc.).

❑ Parts list is complete and agrees with the drawing. Reference designations have been assigned to electrical items.

❑ Drawing and model versions agree with the CAD system.

❏ CAD models will be investigated to uncover problems to the level requested. It may be model and layer naming conventions, status, interferences, and compliance with all company procedures.

❏ Verify that all features are dimensioned and have tolerances.

❏ Notes are complete and unambiguous, and all information contains acceptance and rejection criteria.

❏ Material and finishes are completely specified, including such information as types, classes, temper, and appropriateness for the environments of design.

❏ Datums selected and shown appropriately. Datums are logical, have sufficient size, and are contactable.

❏ Features are located properly with geometric dimensioning and tolerancing, and it has been applied properly:

  ❏ Features located to or from a centerline have been properly defined and dimensioned.

  ❏ Modifier is applied properly, and MMC has not been used just as a default but considered appropriately in the tolerance stackup.

  ❏ Tolerancing is applied appropriately.

  ❏ Data in the feature control frame is properly presented.

❑ Hole size and tolerance are appropriate for standard drill sizes. Hole tolerance has been calculated properly or has positional tolerance and 0.014 used just as a default.

❑ Surfaces having all elements in the same plane (coplanar) have been appropriately toleranced.

❑ Features that require angular orientation have been defined properly.

❑ If painted, is it denoted that dimensions and tolerances apply after plating and prior to painting.

❑ Corner and fillet radius are specified and shown properly.

❑ Deburr and sharp edges note—does it conflict with corner and fillet radii?

❑ Tolerances are not overly stringent and are appropriate for design. Angular dimensions—is the tolerance too restrictive?

❑ Mating parts and interfaces have been verified, and no interferences are present.

❑ There are no other design flaws (interferences) present.

❑ Control drawings have no unnecessary dimensional restrictions.

❑ Assembly drawings have all parts shown with appropriate electrical reference designations.

9

OTHER INFORMATION

DEFINITION OF COMMON TERMS

Understand some of the basic terms so that, when specified, they are used appropriately. If used incorrectly, they will cause confusion.

An example that is frequently seen is, "Inspection to be performed by contractor."

It should have been stated, "Inspection to be performed by procuring activity."

Contractor   A company having a contract with the procuring activity for the design, development, manufacture, maintenance, modification, or supply of items under the terms of the contract.

Datum   A theoretically exact point, axis, or plane derived from its geometric counterpart of a specified datum feature. A datum is the origin from which the location or geometric characteristic of features of a part are established. In simpler terms, a feature on the actual part (datum feature) is used to create the actual datum, and it is the origin from which the dimensions will be inspected. The actual feature is simulated in the processing equipment. For example, it may be the surface plate or machine table that the item is sitting on, or if the datum feature is a hole, then the actual datum may be the pin. Datums should always be selected by functionality and accessibility.

Design Activity   An activity having the responsibility for design of an item.

Original Activity   The activity that had original responsibility for the item.

Current Design Activity   The activity currently having the responsibility for design and maintenance.

Least Material Condition   The condition in which a feature of size contains the least amount of material within the stated limits of size. For example, maximum hole or minimum shaft diameter.

Maximum Material Condition   The condition in which a feature of size contains the maximum amount of material within the stated limits of size. For example, minimum hole diameter or maximum shaft diameter.

Part   One item, or two or more items joined together, that is not normally subject to disassembly without destruction or impairment of the designed use.

Procuring Activity   The contracting activity or procuring activity is the customer.

Regardless of Feature Size   The term used to indicate that a geometric tolerance or datum reference applies at any increment of size of the feature within its size tolerance.

Unit An assembly, or any combination of parts, subassemblies, and assemblies mounted together, normally capable of independent operation in a variety of situations.

Vendor A source from whom a purchased item is obtained.

Table 9.1   Standard and Tolerance

Drill Size	Dec. Equiv.	Drill Size	Dec. Equiv.	Drill Size	Dec. Equiv.	Drill Size	Dec. Equiv.
80	.0135	43	.089	8	.199	X	.397
79	.0145	42	.0935	7	.201	Y	.404
1/64	.0156	3/32	.0938	13/64	.2031	13/32	.4062
78	.016	41	.096	6	.204	Z	.413
77	.018	40	.098	5	.2055	27/64	.4219
76	.020	39	.0995	4	.209	7/16	.4375
75	.021	38	.1015	3	.213	29/64	.4531
74	.0225	37	.104	2	.221	15/32	.4688
73	.024	36	.1065	1	.228	31/64	.4844
72	.025	7/64	.1094	A	.234	1/2	.500
71	.026	35	.110	15/64	.2344	33/64	.5156
70	.028	34	.111	B	.238	17/32	.5312
69	.0292	33	.113	C	.242	35/64	.5469
68	.031	32	.116	D	.246	9/16	.5625
1/32	.0312	31	.120	1/4	.250	37/64	.5781
67	.032	1/8	.125	F	.257	19/32	.5938
66	.033	30	.1285	G	.261	39/64	.6094
65	.035	29	.136	17/64	.2656	5/8	.625
64	.036	28	.1405	H	.266	41/64	.6406
63	.037	9/64	.1406	I	.272	21/32	.6562
62	.038	27	.144	J	.277	43/64	.6719
61	.039	26	.147	K	.281	11/16	.6875
60	.040	25	.1495	9/32	.2812	45/64	.7031
59	.041	24	.152	L	.290	23/32	.7188
58	.042	23	.154	M	.295	47/64	.7344
57	.043	5/32	.1562	19/64	.2969	3/4	.750
56	.0465	22	.157	N	.302	49/64	.7656
3/64	.0469	21	.159	5/16	.3125	25/32	.7812
55	.052	20	.161	O	.316	51/64	.7969
54	.055	19	.166	P	.323	13/16	.8125
53	.0595	18	.1695	21/64	.3281	53/64	.8281
1/16	.0625	11/64	.1719	Q	.332	27/32	.8438
52	.0635	17	.173	R	.339	55/64	.8594
51	.067	16	.177	11/32	.3438	7/8	.875
50	.070	15	.180	S	.348	57/64	.8906
49	.073	14	.182	T	.358	29/32	.9062
48	.076	13	.185	23/64	.3594	59/64	.9219
5/64	.0781	3/16	.1875	U	.368	15/16	.9375
47	.0785	12	.189	3/8	.375	61/64	.9531
46	.081	11	.191	V	.377	31/32	.9688
45	.082	10	.1935	W	.386	63/64	.9844
44	.086	9	.196	25/64	.3906		

Table 9.2   Standard Drill Hole Tolerances

Hole Dia.	Tolerance
.0135–.125	+.004/–.001
.126–.250	+.005/–.001
.251–.500	+.006/–.001
.501–.750	+.008/–.001
.751–1.000	+.010/–.001
1.001–2.000	+.012/–.001

Table 9.3   Stock Thicknesses

Aluminum	Steel	Aluminum	Steel
.010	.0149 (28 GA)	.125	.1196 (11 GA)
.012	.0179 (26 GA)	.160	.1345 (10 GA)
.016	.0239 (24 GA)	.190	.1644 (8 GA)
.020	.0299 (22 GA)	.250	.1793 (7 GA)
.025	.0359 (20 GA)	.312	.25
.032	.0418 (19 GA)	.375	.312
.040	.0478 (18 GA)	.50	.375
.050	.0598 (16 GA)	.75	.50
.063	.0747 (14 GA)	.875	.75
.080	.0897 (13 GA)	1.00	.875
.090	.1046 (12 GA)		1.00

CONCLUSION

By paying attention to the areas that have been shown to cause most drawing and design errors, these problems can be reduced to a more acceptable level. Hopefully, drawing, change activity, production, and inspection costs will be reduced.

With special attention to the following areas, our endeavors will be more rewarding. The following is not a checklist but just a reminder of the more important areas.

- If you are dimensioning to or from a centerline, is it defined?

- If you are dimensioning to or from a centerline, try to visualize it as an axis and not as a crosshair.

- If applicable, is angular orientation required?

- Are you using a maximum material modifier as a default without understanding the impact?

- Does all data contain acceptance and/or rejection criteria? Check your notes especially.

- Do you know what your design and documentation requirements are? Are you following your engineering procedures, or do you have a contractual requirement? If you have a contractual requirement, do you know what is required?

A Final Thought or Two

Always understand all the information that you placed on the drawing. If there is ever any information that you don't understand, don't be afraid to ask. So often I have seen people who don't want to ask because they don't want to look dumb, and when they finally get the courage to ask, they find out that others don't know, either. The others were afraid to ask, just like you. If you have a question, why not ask the engineering design checker? He will love it, especially since he won't have to mark up a misapplication on your drawing.

INDEX

ABOUT THE AUTHOR

The author has over 30 years of experience in the engineering field and has been active in many major defense and commercial programs with some of the top firms in the United States. He has developed projects at Lockheed Martin, Magnavox, General Dynamics, Litton Data Systems, Motorola, and Borg Warner.

His experience includes positions as Staff Engineer, Engineering Electrical and Mechanical Design Analyst, Mechanical Designer, Engineering Procedures Specialist, Customer Contractor Negotiator, and instructor for Technical Data Packages Requirements and Geometric Dimensioning and Tolerancing.